Atwell

Cool Careers in
ENGINEERING

Sally Ride
Science

CONTENTS

Ellen

Dava

Bernard

Oren

Natalie

Wanda

Alexandria

Soojin

Chuck

Oksana

Ayanna

Emir

What Do You Want to Be?

Is being an engineer one of your goals?

The good news is that there are many different paths leading there. The people who work in engineering come from many different backgrounds. But, they all have one thing in common—they're problem solvers. Engineers are behind everything from basketballs and biofuels to laptops and lasers, and from satellites and solar panels to rovers and running shoes, and more.

It's never too soon to think about what you want to be. You probably have lots of things that you like to do—maybe you like doing experiments or drawing pictures. Maybe you like working with numbers or writing stories.

SALLY RIDE
First American Woman in Space

The women and men you're about to meet
found their careers by doing what they love. As
you read this book and do the activities, think
about what you like doing. Then follow your
interests, and see where they take you. You just
might find your career, too.

Reach for the stars!

Sally K Ride

"Being an astronaut isn't just about the science. An astronaut must be a team player."

ELLEN OCHOA

NASA Johnson Space Center

Engineering Smarts

On board the Space Shuttle, Ellen used her engineering smarts to operate the Shuttle's computers and its robot arm. The arm is used to move equipment, satellites, and even astronauts outside the Shuttle. Guess what her favorite space activity was? Like many astronauts, it was floating weightless inside the Shuttle. "It's definitely something you miss back on Earth."

President Ochoa

Ellen didn't think of becoming an astronaut as a girl. She wanted to be president. It's a good thing for NASA that Ellen changed her mind.

Switching Gears

In college, Ellen kept her options open—she switched majors five times! But she always took lots of science classes, and eventually earned a degree in physics. Then, when Ellen was studying electrical engineering, she was bitten by the space bug. She started thinking about becoming an astronaut. Ellen was one of 22 astronaut candidates chosen out of 2,000 men and women who applied. She went on to become the first Hispanic woman to go into space.

An astronaut travels into space to explore our world and beyond. Astronauts come from many different backgrounds, including aviation, engineering, science, and teaching. They use their different skills in space. Ellen controlled the Shuttle's robot arm. Other **astronauts**

* conduct experiments inside the Shuttle.
* release satellites into space.
* study Earth's weather and geology from above the atmosphere.

Edible Engineering

3-2-1 . . . eat! Do a little engineering as you team up with a partner and design an *edible* model of a robotic spacecraft to orbit the Moon. That's right. The only building material you can use is food. Make sure your model has the main parts—an orbiter with a communications antenna to send and receive messages, solar panels to power it, and cameras to photograph the Moon's surface. Make a tasty sketch of your edible spacecraft and present the design to your class. With your teacher's permission, bring the ingredients to class, and then follow your sketch to create it.

Spacercise

Weightlessness can weaken astronauts' bones and muscles. So, it's important for astronauts to exercise in space. Make a list of the kinds of exercises you think would be possible in space. Make a second list of which ones wouldn't be possible. Compare lists with another student, and then create an exercise plan for an astronaut. Discuss what you think would be the best part of spacercise.

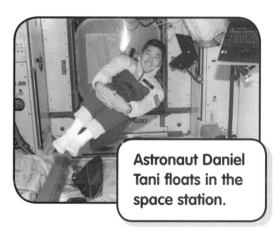

Astronaut Daniel Tani floats in the space station.

Think About

In orbit there is too little gravity to hold things down. How would that affect an experiment like growing soybeans in space? Would the plants still grow upward? What would happen if you tried to pour water on them?

"We should definitely explore Mars together as a world."

DAVA NEWMAN
Massachusetts Institute of Technology

Futuristic Fashions

"We're designing a Martian spacesuit," Dava Newman says. But, unlike the bulky Apollo moonsuits, these will be thin, flexible, and tight—almost a second skin. Slick! But before astronauts make it to Mars, they'll be flying for months. That's why Dava also studies how people adjust to weightlessness.

Step by Step

Dava calls the spacesuit her favorite, but also her "craziest," project. One of her more down-to-Earth projects is a robotic ankle brace for certain stroke victims. Dava studies how people move—in space or at home—and engineers ways to help them. "You get pretty emotional when you see someone walk normally again," she says.

Bodies and Machines

At her Montana high school, Dava was a ski racer. In college, she played basketball big-time. She says that sports steered her toward biomedical engineering. "It really brought together my technological side as well as my enthusiasm for sports."

A biomedical engineer

combines the techniques of engineering with those of medical sciences to improve health care. Dava studies how people's bodies move in weightlessness. Other **biomedical engineers**

* design advanced life-support systems.

* find new ways for people to interact with machines.

* invent medical devices to treat sick people.

What Would U Do?

If you could design an experiment to conduct in space, what would it be?

In a small group, think of a question such as, *What effect does weightlessness have on muscles and bones?* Form a hypothesis. Then design and write up an experiment that would test your hypothesis. What would you measure? How would you analyze the data?

Is It 4 U?

What parts of Dava's job would you most enjoy?

- Teaching
- Designing spacesuits
- Building models
- Helping people with disabilities

Choose one and write a paragraph that explains why.

Suited Up

Imagine you're trekking the Martian terrain. What would your spacesuit look like? Get creative and design one! Make a drawing of your suit and label its features. Include a caption that describes what each feature does and why it's important. Here are some facts about the Red Planet. It has

- an atmosphere that's only 0.1 percent oxygen.
- mostly desert terrain.
- very high mountains and very deep canyons.
- giant dust storms that can blanket the planet.
- temperatures ranging from about 20°C (70°F) to about -153°C (-225°F).

Get feedback from your classmates. How well did you communicate your ideas?

BERNARD AMADEI
Engineers Without Borders-USA

"Who does engineering for the developing world?" Bernard asks. Engineers Without Borders-USA does!

In Nepal, Bernard and volunteers work with villagers to build a system for protecting their water from contamination.

Village Voices

Bernard Amadei's life changed when he visited a small village in Belize that had no easy access to water. Young girls had to haul water from a nearby river instead of going to school. Bernard promised to do something about it. He designed a pump that brought water to the villagers. Now the girls could go to school. It changed all their lives, including Bernard's. "It was the first time I could combine engineering with helping others."

The Engineers That Could

Bernard was inspired. Soon after, he started Engineers Without Borders-USA, an organization that sends engineering students and professionals to places in need. Over 12,000 volunteer engineers have worked on hundreds of projects in over 40 countries. From Afghanistan to Zimbabwe, engineers and villagers work together. They install solar cells to give villages electricity. They dig wells and design irrigation systems. They build schools, bridges, and roads. "Engineering is becoming an agent of positive change for the planet," Bernard says.

A civil engineer designs and constructs public buildings used by many people. Bernard finds creative ways to help people in developing countries. Other **civil engineers**

✷ oversee the construction of skyscrapers and stadiums.

✷ plan highways, railways, and airports.

✷ design bridges, dams, wind turbines, and other structures.

Brick by Brick

In several countries, Engineers Without Borders-USA has programs to teach young people skills that help them earn money. In one program, kids first learn how to collect waste paper and sawdust. Engineers then train the kids to use a hand press to form the waste into round bricks. Villagers buy the bricks and use them for cooking and heating.

Pair up with a classmate. Make a science poster to illustrate the brick-making process, step-by-step. Explain who benefits from each step and why. What's the title of your poster?

What's the Plan?

Imagine you're an engineer like Bernard. Put together a team and choose which project you would like to tackle.

• Build a system to pipe drinking water into homes in Rwanda.
• Use wind turbines to generate electricity for a village in Guatemala.
• Design a water container that uses solar energy to disinfect water for villages in Bangladesh.

Create a proposal for someone who will fund your project. Make sure to explain the need and what you plan to do by answering the *5 Ws—who*, *what*, *when*, *where*, and *why*.

OREN JACOB

Pixar Animation Studios

Movies Move

Imagine making movies on computers, and getting paid for it. Welcome to Oren Jacob's job as a computer graphics engineer at Pixar Animation Studios. He's been technical director on computer-animated films such as *Finding Nemo* and *Toy Story 2*. He oversees the other engineers and high-tech methods used to animate all the creatures and their surroundings. For example, each pelican in *Finding Nemo* had about 50,000 individual feathers. Each feather was created on a computer, and each was programmed to move separately!

Making H_2O Flow

Oren started as an intern at Pixar Animation Studios when he was studying mechanical engineering in college. "I studied how fluids move," Oren says. "Good thing, too, because we had to simulate the way water flows and splashes in almost every single shot in *Finding Nemo*."

Hot Dog!

In elementary school, Oren started learning about and loving computers. He also loved movies. His mother used to take Oren and his friends to the movies and then talk about them over hot dogs and cookies. Let's see—computers plus movies. Now, that adds up to movie history.

Behind Oren is a blowup of a sea anemone drawn by one of the designers of *Finding Nemo*.

A computer graphics engineer

A computer graphics engineer uses computers to create animated characters and scenery. Then they make the characters move and interact with each other. Oren oversees the graphics for animated movies. Other **computer graphics engineers**

✳ produce visual effects for live action movies.

✳ design and create video games.

✳ create and animate special effects or mini-movies for Web sites.

✳ invent new computer animation techniques.

Feather Figures

Each pelican in *Finding Nemo* has 50,000 feathers and each was programmed to move separately by computer graphics engineers. If there were 24 pelicans in the movie, how many total feathers did Oren and his crew have to program? If there were eight people on the crew, how many feathers did each crew member program?

Is It 4 U?

What part of Oren's job would you like, and why?

- Writing computer programs that animate art
- Being part of a movie production
- Getting to see your work on-screen in a theater
- Turning written stories into pictures

Talk about your answers with a classmate.

Movie Homework

Want to learn about computer animation? Oren suggests it's best to study animated movies. After you get an adult's permission, watch an animated movie on your computer or TV. Pause every few minutes. Create a chart to log what you see. Make four columns and label them *Scene*, *Lighting*, *Characters' Expressions* and *Background*. Each time you stop, briefly describe the scene in the first column, then fill in the other columns with your observations.

Check out your answers on page 36.

NATALIE JEREMIJENKO
New York University

Go Fetch

If you see a pack of robotic dogs sniffing a landfill near your home, don't be alarmed. Natalie Jeremijenko and her students are probably there. She likes to buy cheap robotic toy dogs and take them apart. Her students install new brains, new legs, and new noses so e-Fido and pals can trek off-road to search for nasty toxins. Polluters, beware of dog!

Science That's Hard to Miss

Natalie has created lots of cool tech designs. They include a toy spy plane, a kid's pony ride that shakes to recreate an earthquake, and some remote-controlled geese. She has also designed a computer program that prints an image of a slice of tree whenever your printer uses a tree's worth of paper. Green is good! Natalie is famous for dozens of trees that she planted around San Francisco to illustrate environmental effects on tree growth. The common thread is that all her projects present scientific ideas and data in ways that grab your attention.

Crossing Boundaries

So is Natalie an artist or an engineer? Well, in school she jumped between neuroscience, computers, art, and mechanics. "You can ask the same questions from different points of view as a way to develop a better understanding," Natalie says. "It doesn't feel like I'm jumping around at all."

A design engineer can come from different engineering backgrounds, but must be creative. Natalie builds things that present information in new ways. Other **design engineers**

* write multimedia software.
* design sleek, new vehicles.
* create cool electronic devices, such as MP3 players.
* build homes that look cool and save energy.

This robot can learn from the commands its owner sends. Good dog!

Rules of Brainstorming

Brainstorming is an important part of being a design engineer. It's a blast, especially when you stick to these rules.

- **All ideas must be written down or sketched.**
- **All ideas must be respected.** Don't judge or debate. Encourage.
- **There are no bad ideas.** Even ideas that don't seem like winners can jiggle other brains.
- **Don't hold back.** Wild and crazy is good. It's easier to tame a wild idea than to come up with a new one.
- **More, more, more.** The more ideas the better.

Design Makeover

Every design improvement has grown out of an idea. Team up with other students and brainstorm how you could improve the design of something you use every day. It could be the design of your locker, or your lunchbox, or your bike, or . . . you choose! Check out—and follow—the Rules of Brainstorming on the left.

Eye Catching

Natalie uses art and engineering to grab people's attention. Think of a way in which you could use art to bring attention to an environmental problem. Then sketch, paint, or draw your idea.

"I love inventing new technology that will help people."

WANDA GASS

Texas Instruments

She's a Pioneer

Thank you, Wanda Gass. Over twenty years ago, Wanda began working on microchips that today are the brains of some of our snazziest electronics—MP3 players, cell phones, video cameras, and others. The microchips convert the sights and sounds that surround us—voices, video, and music—into digital format. All you have to do is click your mouse or push a button and they can be saved and played back at any time.

Wanda's Way

Wanda also dreams up new products. "I like to try to predict the future—what people will need," Wanda says. Imagine wearing a medical device that monitors your health and tells you what to do in an emergency. At one point, Wanda wanted to be a doctor so she could help people. But she also wanted to be an engineer, like her dad, so she could create cool technology. Today, Wanda does both. That's engineering.

Look What I Can Do

Wanda and a team of engineers started a physics camp for girls. The hands-on camp shows girls that physics is fun and interesting. Campers work with electricity and wire circuits. "I love our work that encourages girls to take on science and engineering," Wanda says.

Wanda doesn't back away from challenges—even those on the water.

An electrical engineer

designs electrical and electronic systems for power plants, satellites, and high-tech products. Wanda designs chips to process digital signals for many electronic devices. Other **electrical engineers**

* develop telephone and cable systems.

* design computers and household appliances.

* create lighting systems for homes and offices.

* monitor electricity to make sure it flows to towns and cities.

Look Around

The list above is just a few of the things electrical engineers do. Find several things at your school or home that an electrical engineer worked on. With a partner, make a list, then share one with your class.

About You

The Explorer's Club at Wanda's high school was a cool science and adventure club. All the members were boys—until Wanda joined. Have you ever been the first to do something?

It's a Process

What would you like to create if you were an engineer? Work as a team to come up with an idea. Follow the *Engineering Design Process* below. Start by *Brainstorming* ideas. Then, *Research, Develop,* and *Choose* your idea. Now, *Create a Drawing*. As a team, present your idea to your class—it's a great way to *Test* your idea.

Engineering Design Process

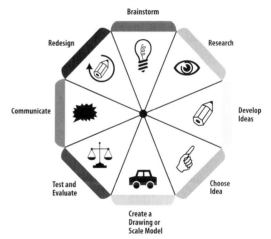

Role Models

Wanda's dad was one of her role models. Who are yours? Write about someone you admire. It can be a family member, friend, or famous person. What have they done that inspires you, and why?

ALEXANDRIA BOEHM

Stanford University

"That's definitely what I enjoy most—being outside doing science."

Environmental engineers develop alternative sources of energy, such as wind turbines.

Diving In

Ali Boehm is never far from water—whether she's working or playing. Growing up in Hawaii, she loved to snorkel, scuba dive, and surf. When the places where she swam started filling with pollution, she knew something had to be done. "I've been an environmentalist since I was really young," she says. "When I was at Caltech, I found out about environmental engineering. It's a way for me to work on problems and still be able to pursue my interest in environmental science."

Into the Wild

Today, Ali and her students visit beaches and take water samples. They use the samples to measure dangerous microorganisms from sewage that might have leaked into the ocean. Their goal is to find sources of pollution and prevent hazards. "I try to get out into the field and do a lot of water sampling with my students." And even though Ali teaches a lot, she still takes time to ride the waves. Hang ten!

An environmental engineer finds ways to protect our natural resources. Ali tries to prevent sewage from ending up in swimming areas. Other **environmental engineers**

✴ build structures that prevent soil erosion.

✴ help design buildings that don't hurt their surroundings.

✴ manage oil spills or toxic waste cleanup.

Career Interests?

As a girl, Ali liked snorkeling and scuba diving. What are some things you like to do? Could they be part of your future career? Talk about your interests with a classmate.

Is It 4 U?

What parts of Ali's job appeal to you? Why?

- Working outside
- Teaching students
- Helping the environment
- Collecting and testing samples
- Measuring microorganisms in seawater

Water Logged

Do you know how much water you use in a day? You might be surprised. Keep a log for one week. Each day, log how much water you use for different things. What's the total number of liters (gallons)? After one week, compare totals with your classmates. Here are some typical amounts of water used.

Washing face or hands	4 liters (about 1 gallon)
Taking 5-minute shower	40 liters (about 10 gallons)
Taking bath	190 liters (about 50 gallons)
Brushing teeth (with water running)	23 liters (about 6 gallons)
Brushing teeth (without water running)	4 liters (about 1 gallon)
Flushing toilet (not low-flow)	11 liters (about 3 gallons)
Flushing toilet (low-flow)	6 liters (about 1½ gallons)

SOOJIN JUN
University of Hawai'i at Mānoa

Pouch Potatoes

When he was younger, Soojin Jun thought he wanted to be a baseball player. In college, he chose a different playing field—food engineering. Today, Soojin cooks up tasty technology for NASA. Soojin has created a plastic and aluminum pouch that is little, light, and heats space food. Someday, when astronauts take the long ride to Mars, their meals might come in this high-tech pouch. A hungry astronaut would just plug the pouch into a heating unit. As the food inside heats up, it's also sterilized. Clever. No one gets sick and the food tastes fresh.

Zap It

Soojin is also working on a biosensor that can detect bacteria in milk, juice, and fresh vegetables. You just dunk the sensor in a sample, and it can "see" the tiny organisms that cause food to spoil. The sensor zaps the food with light, and then "reads" the number of bacteria in the sample. There's no hiding from Soojin!

> "It was fun visiting Kona coffee farmers in Hawai'i to study the effect of environmental factors on coffee quality."

> Soojin studies new ways to process food so it stays fresh longer. Goodbye, stale bread!

A food engineer uses

chemistry, microbiology, and engineering to develop new foods, machinery, and packaging. Soojin designs food packaging for space travel and methods for detecting bacteria. Other **food engineers**

✳ design machines to produce food faster and easier.

✳ develop better methods for packaging and storing food.

✳ advise the designers of new food-processing plants.

✳ discover new ingredients for food.

Factory Direct

Soojin became interested in food engineering on a field trip to a tofu factory. Choose a place you'd like to explore behind-the-scenes, such as a zoo or museum.

Imagine you're a documentary filmmaker and you're going to shoot a movie about the place. Write a short outline of what you'd like to film and why, and who you'd like to interview.

Quiet on the set. Action!

Recipe for Safe Food

What's the food safety score in your kitchen at home? Write down this list of questions and take it home. Investigate, take notes, and then bring your findings back to school. As a class, discuss why each of these is an important question.

- What is the temperature inside your refrigerator?
- How is meat defrosted?
- What happens to meat leftovers?
- Is the cutting board cleaned after raw meat is chopped?
- Do you sample the cookie dough when someone makes cookies with eggs?

Think About

As a boy, Soojin loved reading biographies of engineers. Which engineers inspire you? Are any of them in your family or community? What kinds of work do they do?

"Math is an essential part of my toolbox."

Toy Toolbox

As a boy, Chuck loved to draw and paint. First, he drew what he saw in the world, then what he saw in his mind. In college he received a degree in sculpture, but he wanted to make movable sculptures. So, he went back to school to study mechanical engineering. "Engineering was the discipline that gave me the tools to build the kind of ideas that I had," he says.

CHUCK HOBERMAN

Hoberman Designs

More Than Meets the Eye

Chuck Hoberman likes to make things disappear—almost. You may have played with one of his toy spheres that shrink from the size of a beach ball to the size of an orange. In the 2002 Winter Olympics, when athletes accepted medals on TV, they stood under his giant 22-meter- (72-foot-) wide arch. It opened up like your eye's iris!

Big (and Little) Dreams

Chuck started two companies to put his designs to work in toys, art, and architecture. He's already filed for 17 patents. So what's next? "I want to see my ideas realized in ways that help people," he says. His ideas range from machines the size of molecules to stadium rooftops that retract like garage doors. "The limits are far beyond what I've actually done so far."

Chuck designed the Hoberman Arch for the 2002 Winter Olympics.

An inventor is someone who comes up with new, useful ideas, either from scratch or by combining old ones. Chuck is a sculptor and mechanical engineer who makes unfolding structures. An **inventor** might work in a

✳ research laboratory.

✳ basement workshop.

✳ high-tech company.

✳ hospital.

Chuck's invention "Switch Pitch" changes color in midair.

Thanks!

Each inventor below created something that has made a difference in all our lives. Research one person and her or his invention. Then create an engaging science poster about them. Some things your poster should include—a title, a mini-biography, and an accurate science illustration of the invention. Be sure to tell how the invention made your life better. Think of your poster as a giant thank-you card.

- Mary Anderson
- Patricia Bath
- Douglas Englebart
- Lillian Moller Gilbreth
- José Hernández-Rebollar

Good Boo-Boos

Inventors often make mistakes—lots of them—before they succeed. Think about a time when at first you weren't successful at something. What did you learn from your mistake? Share your story with a classmate.

You're a Rotnevni

As a "rotnevni" you'll be a backward inventor—you'll take things apart. Ask your teacher to give you something you can take apart so you can see how it works. Maybe it's a broken watch or an old alarm clock—after your teacher has cut off the cord. If you pay close attention, and make a sketch, you can even try putting it back together. Enjoy tinkering!

Check out your answers on page 36.

"What makes a good ride is to have a great marriage of technology and creativity. Together they take people into a make-believe world."

OKSANA WALL
Celtic Engineering

Dreams Really Do Come True

When Oksana Wall was 13, she and her family came from Venezuela to visit Walt Disney World®. Oksana wondered who created the amazing rides. She found out they were designed by engineers. Oksana made a decision—that's what she wanted to be. Plus, she wanted to work for Disney. Oksana's dream came true. In her years at Disney, she worked on all kinds of rides—from rocking roller coasters to zooming cars.

Imagine That

Oksana now has her own structural engineering company, and she's still engineering entertainment. Sometimes she and her design team update old theme park rides. They redesign the track, vehicles, and buildings the rides are in—making them stronger, safer, faster, and more fun. She also designs animated props. What are they? "Imagine you're on a ride and something looks like it's just about to fall on you . . . then it stops." Look out!

To dream up fun, Oksana teams up with electrical engineers, mechanical engineers, and architects.

A project engineer designs and manages structures that lots of people use. Oksana designs theme park rides. Other **project engineers**

✷ oversee projects from early sketches to final structures.

✷ plan budgets and schedules for designing and building.

✷ hire and supervise other engineers on a project.

Loopy Design

What does your dream roller coaster look like? What wild things does it do? Write a short description and create a sketch. Then, team up with other students, brainstorm, and draw a design for a really rocking roller coaster. Some design features to think about are the

- number of hills.
- height of each hill.
- shape of each hill.
- number of loops.

As a team, present your coaster sketch to your class.

When it was invented in 1893, the Ferris wheel had 31 cars and each car carried about 70 people. Compare that with one of today's Ferris wheels (left).

Design and Re-design

As a team, you'll assemble a new coaster—not the design you sketched above. This coaster has one requirement—it has to keep a marble in motion for 5 seconds. Use only these supplies.

- Thin cardboard (or stiff paper)
- Tape
- Scissors
- One marble

Like all good design teams, you may need to change your design many times before it works. Try cutting, bending, and folding the materials. And, try attaching your roller coaster to a wall.

> "I want to plop a rover on Mars and have it call back when it finds interesting science."

AYANNA HOWARD
NASA Jet Propulsion Laboratory

Bio a No-Go
When she was in seventh grade, Ayanna Howard saw a TV show called *The Bionic Woman*. It was about a woman who was rebuilt with robotic limbs. The show made Ayanna want to study to be a medical doctor. Then she had to dissect a frog. Ayanna decided she didn't like biology so much! From then on, it was all about robotics. Now Ayanna finds ways for robots to navigate distant lands on their own.

Thoughts for Bots
The trick is getting robots to think like people. Ayanna programs their artificial intelligence to be flexible so they can react to surprises. They can also learn as they go. But engineering is more than just understanding how humans think. "I come up with a beautiful theory," Ayanna says, "and then I get it working on the robot."

Robot Roommates
Ayanna also finds ways for robots to cooperate with people—in space and on Earth. Astronauts may build Moon colonies alongside bots. But back at home, someone who is sick may need a robot to fetch medicine from another room. "Robots are going to become part of our everyday lives," Ayanna says.

Some robots are designed to move like insects.

A robotics engineer programs, designs, or builds machines that can do tasks on their own. Ayanna programs robots to explore Mars and to cooperate with people. Other **robotics engineers**

* design artificial muscles to replace motors and pistons.
* build robots to do dangerous jobs, such as mining and cleaning up toxic waste.
* invent tiny robots that can move like snakes or bugs.

Wall to WALL-E™

Bots can do lots. Write down ten ways you could use robots in your everyday life. Give yourself 5 minutes to complete your list. Compare your list with a classmate's. Then come up with categories—such as transportation or chores—and group the tasks for which you would use a robot.

Go Deeper

Using your list above, pick one idea to explore more. What would your robot do? What would it look like? How would it move, act, and operate? Create a detailed drawing and include a caption that describes its special features and functions. Don't forget to give your bot a name.

Ayanna tinkers with a SnoMote—a robot designed to be like a snowmobile. It may be used to explore the coldest, windiest place on Earth, Antarctica.

Robot Competitions

Would you like to build a robot? Program it? Race it? Many schools and communities have robotics clubs. Check into joining one by asking your teacher or doing some research.

EMIR JOSE MACARI

California State University, Sacramento

Not Too Wet, Not Too Dry

Emir Jose Macari loves to go to the beach and build sand castles. Being a soil engineer comes in handy. "If the soil is too wet, the castle slumps. If it's too dry, it crumbles. A little bit of moisture creates suction that keeps the sand particles together so you can build beautiful sand castles," says Emir. Don't ever compete against a soil engineer in a sandcastle contest!

Dirt Detective

Emir studies how different types of soil behave in earthquakes. When he was working on his Ph.D. in 1985, a huge earthquake struck his native Mexico City, demolishing large parts of the city. What made the earthquake so destructive? Emir investigated the soil for answers. It turns out that Mexico City is built on an old lakebed. When the ground shakes, the waterlogged soil becomes like quicksand. Buildings sink in the wet soil and fall apart. Emir used this knowledge to help engineers design safer buildings. Thanks, Emir, for doing such good work in such a "dirty" job.

"Soil is a very interesting material," Emir says. "Everything we build has to be on top of it."

Emir shows his colleagues and students why the levees failed in New Orleans after Hurricane Katrina.

A soil engineer studies the properties of soil and how well different types of soil hold water. Emir uses his knowledge to help design earthquake-proof buildings. **Soil engineers** also

✳ design plans to clean up soil contaminated by pollution.

✳ use measurements of lunar soil to help design structures that may stand on the Moon one day.

✳ test soil in areas where homes or offices may be built.

✳ design levees that can protect people from floods.

Is It 4 U?

What parts of a soil engineer's job would you like best? Think about it, and then talk to a classmate about what she or he would like about the job.

- Being a "dirt detective" and figuring out the properties of soil
- Working outside to collect soil samples
- Consulting on the design of a building
- Helping city officials to make their communities safer

Can You Dig It?

With your teacher's permission, check out the soil in different areas of your schoolyard. Dig it, think about it, then log it. Create a data table, listing the locations and properties of the soil. How are the samples similar or different?

The major properties to look for are

- **color**—is it mostly black, brown, red, gray, or white?
- **texture**—is it sticky, crumbly, smooth, or gritty?
- **size** of particles—are they small, medium, or large?
- **compactness**—how many centimeters can you poke a pencil into the soil?

About Me

The more you know about yourself, the better you'll be able to plan your future. Start an **About Me Journal** so you can investigate your interests, and scout out your skills and strengths.

Record the date in your journal. Then copy each of the 15 statements below, and write down your responses. Revisit your journal a few times a year to find out how you've changed and grown.

1. *These are things I'd like to do someday.*
   Choose from this list, or create your own.

   - Float weightless in space
   - Operate a robotic arm in space
   - Design theme park rides
   - Create an organization to help people
   - Start a science program for students
   - Use a computer to make a movie
   - Find creative ways to help improve communities
   - Create cool tech designs and toys
   - Help protect natural resources
   - Research ways to keep food safe from bacteria
   - Start my own company
   - Build or program robots

2. *These would be part of the perfect job.*
   Choose from this list, or create your own.

   - Being outdoors
   - Making things
   - Writing
   - Designing a project
   - Observing
   - Being indoors
   - Drawing
   - Investigating
   - Leading others
   - Communicating